Pressed Plants
Naturally

Pressed Plants
Naturally

Bernd Hildebrandt

Hi-Creative

First published in the UK in 2019 by
Hi-Creative Design, Reading, Berkshire, UK

www.hi-creative.co.uk

ISBN (English Edition): 978-1-9162464-0-9
ISBN (German Edition): 978-1-9162464-1-6

10 9 8 7 6 5 4 3 2 1

Pressed plant designs, diagrams and text: Bernd Hildebrandt
Copy-editing: Markus and Martina Hildebrandt
Photos, scans, image editing and design: Markus Hildebrandt

Printed by Swallowtail Print, Drayton Industrial Park, Norwich, UK
www.swallowtailprint.co.uk

Contents

Introduction	7		Winter and Spring Flowers	38
Tools and Accessories	9		Summer and Autumn Flowers	49
Using the Flower Press	11		Foliage	60
What to Press?	13		Unusual Experiments	66
Nectar, Pollen and Sap	16		Conservation of Wild Plants	68
Stem Reduction	18		Taking Pressed Designs Further	69
When to Collect Plant Material	20		Personal Stationery	70
Questions Answered	20		Many Colours …	71
Presenting Pressed Plants	21		Creating Non-floral Pictures	72
Overlapping Stems and Leaves	23		Gallery	82
Note of Caution	24		Glossary	88
Wild Flowers	25		Bibliography	90
Grassy Areas	36		Index	92

The images printed in this book are mostly smaller than the original plants; those that are actual size are indicated (1:1)

Dedication

To the memory of my wife Ilse,
who introduced me to the pressed flower craft
over fifty years ago.

Acknowledgements

My grateful thanks go foremost to my immediate family. Literally from cover to cover this publication was only made possible by my son Markus, through his commitment to this project and his skilled work in presentation. In recent years he created an archive of over twelve-hundred scans of my pressed plant designs, the original pictures being sold at charity events as greeting cards. In all this we had the invaluable practical support of my daughter Martina, an artist in her own right, who also engaged herself effectively over many years in the charitable application of all my work with pressed plants.

I thank my neighbour Charles Langham, for his interest and a generous supply of material from his garden and allotment.

My thanks go to the many people all over the world, known and unknown, who expressed their admiration for this work and gave me encouragement. This can be summed up in one quote: "You have given pleasure to hundreds of people – don't forget that!"

Introduction

This book offers an alternative presentation of pressed plant materials to those found in other craft books and magazine articles on the subject. It describes and illustrates a different approach to processing materials that nature provides, for surprising and pleasing end results.

Throughout the history of art and craft, plants have their firm place: they can be observed in Roman mosaics, in the earliest illuminated manuscripts, in stained glass, ancient stonework, medieval woodcarving, and paintings of all ages. Everywhere, even in the most stylised forms, plants are depicted as a whole, as objects of beauty in their own right. All parts of a plant – stems, leaves, buds, flowers, fruits or seedheads, even roots – inspire the artist and artisan. But rarely would you find a flower from one plant and leaves from another put together and presented as a new composition. A curious practice of mixing came about only in relatively recent times, creating an entirely new genre in the application of plant material in art and craft. What we have is a creative Victorian pastime that has undergone very little change.

To this day, the usage of pressed flowers and other parts of plants in the design of pictures, without any regard to botanical accuracy, represents this once very popular activity. In craft books the reader is guided towards designs that often result in elaborate, attractive, but completely unnatural presentations. And because instructions are often given to press only blooms from certain plants, stalks from others and leaves from again an entirely different source, the unique character of a plant is not taken into account. For instance, I remember a charming picture of pressed pansies, but the two maple leaves, which formed the base of the design, illustrated perfectly the common practice. Why not use the plant's own, very attractive leaves? Maybe because they were not on the list of suitable material.

Cow Parsley
Anthriscus sylvestris
'Ravenswing'

From book to book the process of pressed flower design varies so little that you get the impression there is only one way to approach the subject. Basically: "Books on pressed flowers are derived from old books on pressed flowers that are derived from old books on pressed flowers... This is called tradition." (based on a humorous text about educational books by the writer Erich Kästner, 1899–1974). Tradition is important, but even established experts may welcome a new initiative.

The alternative approach to pressed floral art described here suggests that you put aside the traditional books on the subject, resist the colourful grand designs, and start afresh with an open mind for stimulating and rewarding experiments. So, references to the usual practices are kept to a minimum, although some points of important differences are made. The concept advocated here is based on the premise that a plant is in harmony with itself, and that design should therefore aim for botanical accuracy wherever possible.

Daffodil *Narcissus* and
Grape Hyacinth *Muscari*

That said, there remains plenty of scope for artistic input, only the emphasis is shifted towards incorporating most of a plant's own characteristics. Sometimes, however, you have to forego a feature altogether, or in other cases you can imitate an important part of a flower by taking it from another plant without compromising the natural appearance. Only an expert may notice the modification, as for example, the daffodil (left) with stand-in snowdrop leaves, and the rose (right) that has a whorl of stamens borrowed from another flower (see pp40 and 54 respectively).

Without contradicting the main purpose of this book, I have included examples of using plant material in non-floral art (see pp72–81 and the colour chart p71).

Rose *Rosa*

Tools and Accessories

A flower press is top of the list. You can buy one or make your own. A bought press may need to be adapted for best results. I now have six presses in use all year round, because at times one is not enough to accommodate material that is available for a limited time only. Ready-made presses should neither be too small nor too large. Sizes of approximately 18 x 18 cm or 18 x 23 cm are good. The ideal size for a home-made press is probably 15 x 21 cm, because it can accommodate inlays of A4 card that is cut or folded to A5 with the corners cut off. Make sure that the press is made of 9 mm thick plywood, as other material such as MDF is too rigid. For a satisfactory pressing process the press needs to flex to maintain pressure for longer, otherwise the plants may shrink and distort as they dry. The length of the bolts determine the amount of plant material a press can take: I find 60 mm is just right and not too long. Metal washers under the wing nuts are a good idea, because they prevent damage to the plywood cover over time. Also, nothing is more annoying than bolts slipping out every time you open the press. Mine are therefore firmly glued into the base.

Traditionally, sheets of blotting paper are sandwiched between corrugated cardboard in the press. However, these can cause lines across pressed material due to the uneven surface and therefore pressure. For best results, I use sheets of thin card of 160 g/m^2 that have good absorption quality. And instead of using corrugated cardboard, place your sheets of this thin card either between bought recycled greyboard or other absorbent non-corrugated cardboard from e.g. writing pads or boxes cut to size. A good thickness is about 1 to 1.5 mm. This arrangement of thin card and greyboard does not only give flatter surfaces for the plant material, it also increases the amount the press can accommodate. I only use additional corrugated cardboard to act as cushions in the bottom and top, and maybe one layer in the middle of the press, to even out the pressure.

Because I highly recommend a press put together in this way, I would consider it a good investment to buy a pack of 250 sheets of A4 card of 160 g/m^2. For storage of pressed material, shallow A4 size cardboard boxes that hold 30 to 40 of these sheets are easier to handle than deeper ones – the material stays flat, dry, and retains its colour because it is kept in the dark.

1) Greyboard 1–1.5 mm, card 160 g/m^2, and corrugated cardboard (A5, corners cut)
2) Flat washer M6, zinc plated
3) Wing nut M6, zinc plated
4) Machine screw pan head slotted, M6 x 60 mm, zinc plated
5) Plywood 9 x 150 x 210 mm

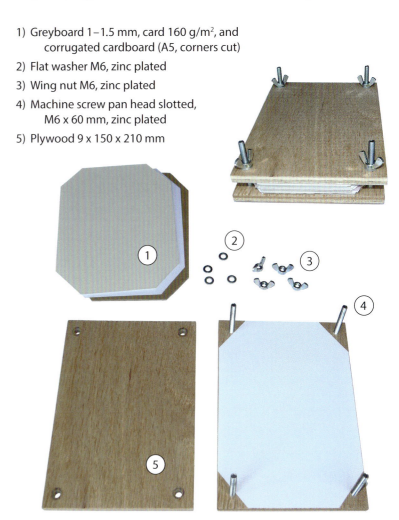

Once you are set up, you can re-use the card and greyboard sheets in your press again and again.

In conjunction with the flower press, a small pile of old books is most useful. They can make a great addition in the process of pressing plants.

Throughout, from pressing the material to its final application, a sharp craft knife with a pointed blade is needed.

A pair of sharp medium-sized scissors should ideally be set aside and only used for working with pressed plant material.

A small hardback book that has outlived its original purpose becomes very useful on walks in the countryside. Plants placed between the pages of the book, which is then held closed with a rubber band, is the best way of transporting them until they can be transferred into the press. The aim is not for the plant to be held fresh, but to have it already available in a semi-pressed position. This method can only be improved with the actual flower press at hand, but this is not always practical.

We now come to the basic items needed for the presentation of pressed plant material. These can include greeting card blanks (with straight edges if you want to cover them), sheets of white or pastel coloured card, or any suitable object you may want to decorate.

A good clear-drying liquid paper glue should be used to hold the plant material in place.

Also needed are: a washable sheet of firm plastic of about A4 size (ideally white), a medium-sized soft artist's paintbrush (round, no. 4 or 5) and a pair of round tip tweezers as used by stamp collectors. For folding pre-scored cards or scoring purposes a bone folder is a useful additional tool.

Certain work can be protected with a transparent polyester sheet, or good quality self-adhesive film.

Designs can also be framed behind a sheet of glass or acrylic. What suits or sets off your work best is a matter of choice.

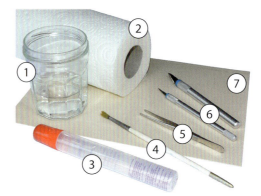

1) Jar of water
2) Kitchen towel
3) Paper glue pen
4) Artist's paintbrush
5) Tweezers
6) Craft knives
7) Firm plastic sheet

What is not needed are watercolours or any other paint for touching up. Adding colour to a display always shows, spoiling the plant's natural attractiveness. Material that does not work on its own is better left out. Sometimes, however, a few pencil lines added to a pressed plant picture (see p26) can give an interesting effect.

Creeping Cinquefoil *Potentilla reptans*

Using the Flower Press

Contrary to the often given advice to keep material in the press for long periods of up to six weeks, I aim for the shortest possible time, i.e. on average only six days. I found that this preserves the material in the best possible condition. How is such a short time achieved? In winter, when the central heating is on, I place the press, standing on its side, directly on top of a radiator, but any place with dry warmth suits well. Although the colder times of the year do not coincide with the season when flowers are most plentiful, the winter provides the most resilient of wild flowers, in gardens often regarded as weeds. The spring brings blooms and fresh green that, because of high moisture content, benefit from a quick pressing procedure. There is also the all year round range of flowers from florists and other retailers to consider.

As soon as it becomes warm enough to dispense with indoor heating, achieving short pressing periods becomes more difficult. Just at the time when you may wish to use your press the most, even in the hottest summer days, the drying in the press slows down to about double the time I ideally aim at. Leaving the press for long periods in blazing sunshine heats it up, but does more harm than good. I suspect that the humidity in the press, especially in the inevitable cooling off period that follows a baking, actually increases. A more even temperature is therefore preferable. As the shortest possible time in the press is still the all-important aim, not least because some plants flower only over a very short period, the following routine seems to overcome the problem. Because the main purpose of the press is to make the material evenly flat, there is actually no benefit in leaving it in the press until fully dry. So, after about a week, the plants can be transferred to an old book (slightly weighed down) or a pile of newspapers (but not glossy magazines) to complete the drying process. It is crucial that during the transfer the material remains undisturbed between the sheets that were

Wild Carrot *Daucus carota*
Flowers, stems and leaves after pressing

originally placed in the press. Therefore additional sheets of card are needed to replace those taken from the press. It is a good idea to give the greyboard a few hours airing, instead of returning it to the press straight away.

Most important is the actual pressing procedure in the press. Ideally the material placed in each layer should be of the same type, in order to react uniformly to the pressing. For example, stalks, leaves, buds and flowers of a plant are often of different thickness. If grouped together on a sheet, the thinnest ones may shrivel up, because the pressure does not reach them. If the wing nuts are over-tightened after fresh plants have been placed in the press, the plants can actually be damaged; when moisture is squeezed out too quickly, colour can be lost. It is therefore advisable to initially apply only gentle pressure, i.e. stop turning the wing nuts when resistance is felt. After about 30 minutes they can already be given a few more turns and fully tightened after several hours. But if you forget to follow this procedure of gradual tightening, the material may have been spoiled by the time you remember it again. By then the plant has not only wilted, but shrunk out of shape. Therefore, pressing plants is not a casual activity, but requires some routine and discipline, time and patience.

Silverweed
Potentilla anserina (1:1)

Wild Carrot *Daucus carota* (1:1)

What to Press?

If you have read elsewhere what is considered useful material, with the added advice to press only certain parts from one plant, my advice is different. Always look at a plant as a whole. Only by preserving as many characteristics as possible will you do the beauty of a plant justice. And this is the aim here. If necessary, see how a plant may be divided into parts that can be pressed separately and successfully reassembled again afterwards. A pressed bloom for which you have no matching leaves and stems/stalks may stay in storage forever. Experiment with plants of various types and sizes, and aim for what will look natural. Often a single specimen displayed botanically as true as possible receives more admiration than a grand design. Have faith in simplicity.

Following this principle, I must have made thousands of pressed flower pictures and other displays for charitable causes over the years. The favourable responses received, I credit entirely to the way nature is preserved in designs that follow the natural structure and habit of a plant. And although some plants can initially look unsuitable or unattractive, they may nevertheless become a pleasant surprise. Others, which have the appearance of good candidates for your press, may fail. For example, I found the colourful Peruvian lily irresistible, only to find,

Peruvian Lily
Alstroemeria

13

after having pressed the blooms reasonably well, that retaining the green in the foliage presented quite a challenge, something that is relatively rare with greens. Nevertheless, after trial and error, a short pressing process of a selection of small to medium-sized leaves of the deepest green brought some good results. Definitely a winter candidate bought from a florist as cut flowers, usually known as *Alstroemeria*.

Unsuitable are the blues of the bellflowers (*Campanula*). In the pressing process the colour disappears, leaving a near transparent tissue. I have not given up on sweet peas, although even with the strongest colours the results are poor. As with the Siberian iris, pressed sweet pea flowers have short-lived colour retention in comparison with the performance of most other flowers. A marked deterioration of colour becomes apparent already after a few months.

I sometimes press just individual bigger petals from a variety of blooms because they provide, together with other surplus material, a large palette of colours and shapes for non-floral art (see pp71–81).

Having used around two-hundred plants successfully, I need to point out that each required individual attention, especially because the way in which a fresh plant is positioned in the press determines its later application in a display. In fact, the designing aspect of presenting pressed plants starts here.

Many wild flowers, garden flowers, and material from shrubs and trees make up my stock. And experimenting with new plants provides an ongoing interest and challenge.

In later sections, explanations and tips are given on selected examples. More plants can be identified from the illustrations. Wherever possible, I have used the common name of plants first, and then the botanical name to help identification or if no recognisable common names are available.

Siberian Iris
Iris sibirica

Siberian Iris (faded)
Iris sibirica

Sweet Pea
Lathyrus odoratus

Sweet Pea (faded)
Lathyrus odoratus

Unexpected Results Although not all tries turn out to be satisfactory, some bring unexpected results. I once found some wall lettuce hugging a high garden wall in a London street and removed some of the shiny young leaves and a couple of stems without endangering the plants. When pressed, the material turned to a nearly black green, actually shown here as a black silhouette design (see also p69).

Wall Lettuce
Mycelis muralis
(black silhouette)

Clematis

Common Ash *Fraxinus excelsior*

Common Ash *Fraxinus excelsior* (winged fruit clusters)

Other materials that turn almost black are the young leaves and the freshly formed winged fruit clusters of the ash, young walnut leaves, the leaves of many *Clematis* varieties and elder (*Sambucus nigra*) 'Black Lace' (p87), to name just a few.

Nectar, Pollen and Sap

It took me a while to find out why pressing flowers like daffodils, gentians and periwinkles all turned out with unsightly brown patches. Following a hunch, I separated a periwinkle corolla tube from the calyx (see Glossary, p88), sliced the tube open, took out the style and stamens and dabbed off the sticky nectar with kitchen towel. This solved the problem; and with daffodils (p40) and gentians (p58) the results were the same. So, nectar and pollen can have an adverse effect when pressing flowers. As a remedy it is sometimes sufficient to take out stamens and style through the corolla tube/corona with tweezers, and to absorb the nectar by inserting a cotton bud. If flowers with long tubes such as verbenas and primulas are to be pressed, it is best to pull the tube out of the calyx and then cut it off as shown (1) in the diagram. By using this treatment, the flowers not only look better, they are also easier to lift off the pressing sheet, because the nectar can act like glue. When you have many small flowers in clusters, as with verbenas, if the tubes are not shortened they would get in the way when reassembling the plant.

When preparing a passion flower (see p50), the centre of the flower will be sticky with nectar and so will be your fingers. It is important to dry off the flower by placing it between two sheets of kitchen towel and squeezing its centre between thumb and forefinger. The process may need to be repeated. If this is not done, a fairly large part of the petals and corona will not only discolour brown, but will also be stuck down firmly on the sheet in the press.

Sometimes, fleshy stems of plants will need splitting before pressing (e.g. daffodil, tulip), and may be moist and sticky with sap on the cut side. If not dabbed dry, the pressed material will not lift off the sheet cleanly; it is not always possible to prise it away without causing damage. I would generally advise to dry off any material that is wet to the touch from nectar or sap. But, contrary to common belief, a bit of moisture from dew or rainwater does no harm.

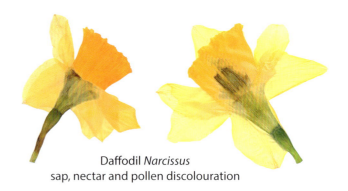

Daffodil *Narcissus*
sap, nectar and pollen discolouration

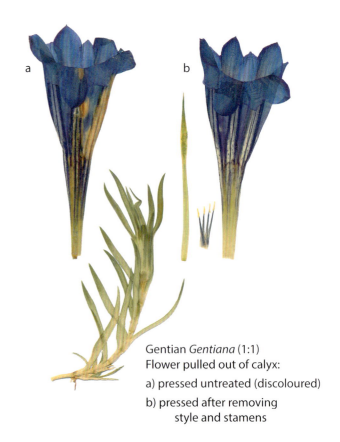

a   b

Gentian *Gentiana* (1:1)
Flower pulled out of calyx:

a) pressed untreated (discoloured)

b) pressed after removing
style and stamens

Nectar
discolouration

Greater Periwinkle
Vinca major

Verbena (annual)
Verbena × hybrida (1:1)

Stem Reduction

Sometimes stems are too hard and/or too thick to press whole and need to be reduced. When attempting this, you will need to work with a sharp, ideally flat-handled, craft knife. Care and patience are called for and a few trials are advisable. Foremost, place the material on a flat surface and hold it down. As you will work the knife towards you in most cases rather than away, the greatest care should be taken.

The material may require additional attention after pressing. Dry stems can often be reduced further by using fine-grit P120 sandpaper instead of the knife.

c) If a robust, not necessarily thick stem needs a flattened side to become suitable for pressing, you have to decide if, for instance, a flower on a curved stem is to be used facing left or right, i.e. the back of the stem is the side on which you shave off as required. If not sure at this stage, see that you have enough material facing both ways to increase your options.

a) Thick hollow soft stems can often be pressed whole, e.g. dandelion. Others, although hollow and relatively soft, benefit from splitting into two right along their length. This can easily be done and you get effectively two stems out of one, like in the case of daffodils, giving you a choice of material later.

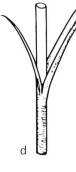

d) Sometimes, woody twigs like those of a maple or larch have no use whole. You may only need to cut a length of bark from one side. This may often allow you to prise out the wooden centre, giving you another length of bark for the press. The result is material that has the appearance of branches of various widths.

b) Stems that are solid with a wide diameter can provide two, three, or even four usable parts for the press. Long, fleshy tulip stems can be treated in this way, as can those of the hellebore. But if there are leaves growing directly off the stem, they need to be detached first, pressed separately and used in the most natural way later. Shorter, multiple stem sections for a display of *Hydrangea* and peony can also be gained through this stem division.

e) A related issue where some reduction or removal of material is needed, arises from the structure of leaves, where the midrib towards the stalk and the stalk itself is much thicker than the rest of the leaf. And regarding roses, the small prickles along the underside of the main vein of a compound leaf are best shaved off before pressing.

Horse-chestnut *Aesculus hippocastanum* (1:1)

When to Collect Plant Material

Collecting plant material for the press is not always straight-forward, and too much emphasis can be placed on timing and the weather. If I would always wait for the so-called ideal conditions, i.e. sunshine and at least 24 hours of dry weather, I would never get much material together. On a wet winter's day I may come across some good rosettes of shepherd's purse or hairy bittercress hugging the soggy ground. I take them in, carefully wash off any soil from the leaves and dry them as best as I can between two sheets of kitchen towel before placing them into the press. Dandelion leaves from the roadside need at times the same treatment. In a quick-drying pressing process the dampness in all these examples does no harm.

The small early-flowering crocus varieties are most suitable for pressed floral art. Without sunshine a crocus will stay tightly closed for days on end and may then open just for a short time when you are not about. But when picked in its closed state, a few at a time, and brought into the warmth of a room, you can watch them as they open. With the press ready, you have the advantage of using the flowers at different stages of opening; this allows later for a more interesting design.

Throughout the year it is always the plant's habit that is the important factor to accommodate. A passion flower completes its flowering cycle often in a single day; daisy flowers, as do some others, close at night and open again when conditions are right.

> *Red poppy petals,*
> *after a summer's day rain,*
> *soon are blown away.*

In this haiku I imply that if I wish to use a rain-heavy poppy flower, I commit myself to a delicate towelling off procedure. Sunshine and dry conditions are certainly helpful, but ultimately trial and error in varied conditions are the order of the day.

Questions Answered

In my experience two questions come up most frequently:

1. How do you retain such good colours? There is no secret about this. Plants are chosen for their suitability in the first place. Experiments that fail in several attempts will eventually not be repeated. Also, attention to detail in the pressing process plays its part. That is why I consider, for instance, all procedures that help to even out the varying thickness of materials as most important. This is also the reason why I have not mentioned any alternative pressing methods, because they do not give me any, or only very limited, control over the desired outcome.

2. Do the colours fade? I used to find this often asked question a little bit irritating, but once you expect it, it is actually quite amusing. Yes, colours fade. Most things we surround ourselves with, fade over time. In relation to plants, we think nothing of it, when we discard a wilted bunch of flowers. We accept that a flowering pot plant may not be worth keeping after the flowers have gone. The concern over pressed flowers therefore defies logic. Yes, they fade, unless you pack them away, never to see the light of day again.

Completed cards covered with self-adhesive transparent film benefit, I have found, from a further period between the pages of a weighed down book. It gives the card and cover extra time to bond together, making the whole more durable, including the colour retention.

On the whole, displayed out of direct sunlight, pressed plants change little over a year or two. When, after thirty plus years a picture of pansies, made for me by my wife, had turned a creamy colour on its velvety black background, I still saw no reason to part with it.

Presenting Pressed Plants

When you have all your pressed material together, you can make your design. With the help of tweezers, loosely assemble the separate parts of a plant or group of plants on a sheet of paper the same size as your chosen background for the display. Once satisfied with the position of the assembled parts, take them bit by bit, starting from the back of the design, and glue them into their final positions. As you make your picture, your original layout will change slightly. Don't expect it to match exactly what you intended; adjustments will invariably have to be made as you progress. Holding each part underside up with tweezers on a washable plastic sheet, apply a small quantity of clear liquid paper glue thinly with a soft paintbrush, in most cases to the whole underside of the plant part, and then stick into place. Use clean sections of your plastic sheet as you go along, and wash it under a tap occasionally. The brush needs to be rinsed from time to time in a jar of water; a little bit of extra water on the brush does no harm – it keeps the glue liquid for longer. Although it dries clean and clear, care needs to be taken not to get glue on the face side of the material. If it does happen, it can be dabbed off with a clean cotton cloth, which you also use for gently pressing down the parts onto the card. Once familiar with the process, you may prefer to press down with the palm of your hand.

There may be occasions when you want to check on a living plant or consult a book, whether leaves, for instance, are arranged alternate or opposite, or to reassure yourself on any other detail. I once pressed fifteen individual flower petals of the St John's wort shrub 'Rowallane' (p52), intending to reassemble them into three flowers. When I started this, I used six petals for the first flower and then realised that I had not enough petals left for the others. I checked on the bush, and as is common with so many flowers it actually had five petals. So, when in doubt, it is always good to check.

Wild flower composition using a card with oval border

Some pressed flower petals are stronger than they look; others, especially large ones, can be so delicate, that a touch of glue can ruin them. In the latter case, just apply a bit of glue to the centre of the flower and place it in position on the chosen background. With the centre fixed, lift each petal in turn by blowing gently under it and have your brush ready to apply glue to the exposed area. When you stop blowing, the petal falls back into place (see diagram).

Sticking down delicate petals
(e.g. pansy)

Sticking down the whole petal in this way is my recommended practice. It becomes especially important when the finished work is to be covered by a self-adhesive transparent film. If you try to cover with film a flower that has not been glued into place, or whose petals are only partially fixed to its backing, the material will literally jump up.

The same principle is applicable to leaves. For instance, a stem of the very delicate maidenhair fern (p84) is glued initially just down the centre and then each leaf treated individually with patience and care.

Wild flower composition

Overlapping Stems and Leaves

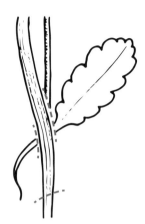

Overlapping material
cut away and a clean
cut end

In many compositions long stems and linear leaves will tend to overlap and/or bunch up. I therefore often cut away material that is not visible because it is covered by something else. A stem cut away at a suitable angle can, if desired, reappear further along. Grass stems and blades can build up to quite a thickness near the base of a design. These stems may look slender, but are hard and do not react to pressing. They should be cut off as soon as they meet blades or leaves of other plants. Blades that overlap can be sorted out in a similar way. For this reason I do not glue any material all the way down straight away, but only once unnecessary parts have been dealt with. Finally, any rough edges at the base can be neatened.

Stem in front of
or behind a leaf

If a flower stem is flat, a leaf can be laid over it without any further attention. However, if a stem is thicker, it will buckle the leaf. It is then best to remove the stem from under the leaf by cutting it at the leaf entry and exit points. Depending on the stem, it can also be cut off above the leaf and butted on again where it would naturally reappear beneath the leaf. This issue does not arise if a stem is placed over a leaf.

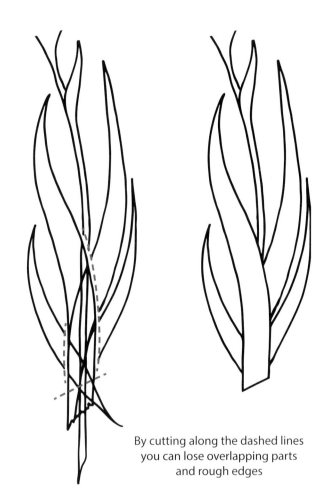

By cutting along the dashed lines
you can lose overlapping parts
and rough edges

Note of Caution

If I look in a book on herbs and medicinal plants I expect to find an account on toxic properties that can be beneficial if applied correctly. However, this information, especially on poisonous plants or plants that can harm in other ways, is sometimes missing or arbitrarily treated in wild flower field guides and books on flowers in general. The same observation applies to seed and plant catalogues and to plant labels.

As this is a craft book and not a botanical study aid nor a guide on how to create a herbarium, a general caution has to suffice. When I therefore mention, that daffodils are harmful if eaten, that hellebores are poisonous in all parts, that the here shown *Laburnum* is highly toxic, I give only examples.

When pressing plants there should always be an awareness that all material, both fresh and dried, should only be used for the purpose intended here, dealt with in a place away from food preparation areas, and be kept out of reach of small children.

Handled with a measure of caution, there should be no need to avoid plants in the pressed flower craft because of the presence of toxic or otherwise harmful substances, like skin-irritating properties. For instance, the pretty meadow buttercups (*Ranunculus acris*) can give you skin rashes and blisters, but become harmless in the pressing process, i.e. when dry. The stinging nettle stings only at the time of picking. Once you have placed it between the pages of an old book as recommended (p30), the plant becomes harmless. With the proper precaution at the time of gathering, there is no need to avoid this plant because of its defensive habit.

When handling plants known to be very toxic, it is advisable to wear disposable gloves.

Remember, as with any form of gardening, to wash your hands thoroughly after handling plants.

Laburnum

Wild Flowers

As all our treasured garden blooms are derived from wild-growing plants, I now give our wild flora its due attention. Some wild flowers are best displayed in groups, just as they can be found together at the wayside or the grassy strips along the pavements of urban streets. These green verges often hold a surprisingly large variety of plants, by no means just grasses. Within a few paces a dozen or more different plants can be found amongst the grass. What you pick here for the flower press does no harm, as the verges are mowed anyway, but please see Conservation of Wild Plants, p68. There are many issues like parked cars that do the often irreversible damage.

Red Campion *Silene dioica* and
Broad-leaved Dock *Rumex obtusifolius*

With plants grouped together from a common environment, nothing will really look out of place, as famously shown by Albrecht Dürer in his watercolour *Das große Rasenstück* of 1503. The composition above shows such a group, which includes meadow grass, hawkweed, daisy, clover and shepherd's purse, all very common plants throughout Europe and further afield. Most wild flowers also make a good show as single specimens.

Dandelion   This has always been one of my favourites, an example of a most resilient plant. We find it in abundance standing tall and robust in a meadow or short but equally robust between paving slabs or gravel. The leaves provide interesting variety. So, care has to be taken in pressing enough leaves of the same shape for a presentation. It would look most odd if you mix the different characteristics. Belonging to the large group of composites makes the dandelion flower difficult to press full-face, but this should not discourage you to try. Don't go for the biggest. Peel away the downwards-curled green bracts underneath the calyx before cutting off the stem as close to the receptacle as possible. Pushing the flower gently face down on a flat surface, with your index finger over the raised centre, gives some idea of the end result before placing it in the press. Buds, and flowers that are not fully open, can be sliced in half from the curled bracts upward, then pressed and displayed in profile. Here the success rate is much greater. Stems should be pressed separately. Choose the more slender ones with a pinkish-green colour for added contrast.

Without any real drawing skills you can add some pencil outlines by placing a leaf where you want it shown and tracing it. Adding the familiar "dandelion clock" needs a bit more practice. This method of adding a drawn leaf to a design can be used for various displays, especially those where a real leaf would crowd the picture. In most cases these additions are not necessary or desirable, but can sometimes be very effective.

Dandelion leaves vary greatly in shape

Dandelion *Taraxacum* (with added pencil detail)

26

Hawkweed Closely related to the dandelion, hawkweed is a complex genus of plants with hundreds of yellow-flowering variations under the botanical name *Hieracium*. An exception is the distinctive orange hawkweed, commonly known as fox-and-cubs (*Pilosella aurantiaca*, formerly *Hieracium aurantiacum*) (see Gallery, p83).

Attempts to distinguish between the many hawkweeds is not easy and in the context of pressed flowers not really necessary. The availability of these plants and usefulness in a wild flower composition is what counts.

* Right: Hawkweed with Hedgerow Cranesbill *Geranium pyrenaicum*

Poppy The common poppy and its garden relatives are tricky candidates for the press as the flowers are so delicate. I found them most attractive in displays if applied in the right way. A few tips may help when handling this interesting plant, which sheds its petals often after flowering for only one day. I cut the flower with a length of stem off the plant, which keeps on producing new growth throughout the summer. Smaller flowers are often more useful than the larger ones. Growing from seed in somewhat overcrowded containers or pots produces many small plants. Leaves and also some nodding buds on their curved stems are gathered in the same way, i.e. a few at a time from different parts of the plant. From a larger flower it is best to carefully nip out the distinct ovary. For all flowers I leave the delicate petals in the position they happen to place themselves on the sheet for the press.

Poppies look best in profile, but a full-face view works sometimes with smaller flowers, where the four petals do not overlap too much. Buds are encased in two sepals that drop off when the flower has fully opened. With a knife I cut a whole bud into two halves from stem to tip and press only the section with its stem attached.

Hawkweed*
Hieracium

Iceland Poppy
Papaver nudicaule

Common Poppy
Papaver rhoeas

California Poppy
Eschscholzia californica

27

Cornflower and Sunflower   A single flowerhead of these composites has many florets enclosed by overlapping, scale-like, green bracts. The ray florets should be pulled out and pressed individually. Only a few are needed to depict the flower in profile on paper. As the lower, globular part of the flowerhead is formed by tough bracts, it is not suitable for pressing. Even cut in half, with the exposed inside of the receptacle removed, the desired flatness cannot be achieved. I therefore use buds with their softer bracts instead. Halved, and with the content cut away leaving just the shell of outer bracts, they actually come quite close to the size of a mature flowerhead when pressed. The same principle applies to the perennial cornflower, and also to the stiff sunflower and the Jerusalem artichoke which are both very similar in appearance.

For full-face depictions of the cornflower (*Centaurea cyanus*) (an annual, shown p33) the issue of bracts does not arise, but then you do have to press the tiny disc florets as well. Wherever possible, the leaves, buds and stems should be pressed as well as the flowers. Stems need splitting and the stem-hugging furry leaves are best taken off, pressed separately, and later put back as the design requires. Very small buds, just appearing between leaves, add to a life-like display.

Perennial Cornflower
Centaurea montana

Stiff Sunflower
Helianthus rigidus

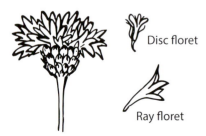

Disc floret

Ray floret

Cornflower florets pulled out
of the receptacle

Bud halved and used as a
stand-in for its mature form

Jerusalem Artichoke
Helianthus tuberosus

Corn Marigold Because of its invasive habit, the corn marigold, like the corncockle (*Agrostemma githago*), has more or less been eradicated in agriculture as an unwelcome weed. However, it can be grown easily in the garden from seed. As with all composites you might wish to press, the receptacle at the end of the stem can be prominent and needs to be carefully reduced by taking off as much as possible, but obviously not to an extent that the flowerhead falls apart. If, in the pressing process, an outer ray floret is lost and an unwelcome gap appears, this can be repaired by using one of a similar size from another slightly damaged flowerhead retained for just such a purpose.

Very similar in appearance and growing habit is the crown daisy. It could easily be mistaken for the corn marigold, however its leaves are distinctly different. It is pressed in the same manner.

Corn Marigold
Glebionis segetum

Crown Daisy
Glebionis coronaria

Flowerhead with a gap

Glue in outer ray florets on the underside

Repaired flower (1:1)

29

Stinging Nettle This can be an interesting pressed plant in its own right, but combined with other wild flowers its structure provides a welcome contrast. As stinging nettles can usually be found in abundance, you can be choosy. I take only the tip of a slender plant, down to maybe four sets of leaves, and preferably with clusters of male flowers of these dioecious plants. It is a good idea to press extra leaves of the same size from the same patch of plants, because they can vary in colour and shape from site to site. Sometimes thicker stems need to be reduced, i.e. cut in half (see p18). It is unavoidable that some parts of the delicate flower clusters get detached. They can be pressed separately and put back into position later.

Freshly gathered material is best placed in an old book until it can be transferred into the press (see Note of Caution, p24). At the transfer stage some arranging of leaves and tangled flowers can be made.

Wild flower compositions that include the stinging nettle

Stinging Nettle *Urtica dioica*

Bird's-foot Trefoil   As with sweet peas (p14), most flowers of the large pea family do not retain their colours well when pressed, including lupins (*Lupinus*), everlasting peas (*Lathyrus latifolius*) and small vetches (*Vicia*). Exceptions are some of the yellow-flowering plants. Together with the meadow vetchling (*Lathyrus pratensis*), the bird's-foot trefoil (just one of its many names) is most valuable in a wild flower composition. Young, not too mature seed pods add interest, but large seed pods or ripened ones are not really suitable for the press, unless you embark on the fiddly process of cutting away the back of each pod and extracting the seeds.

Sometimes I use vetches (especially the small early spring growth) like hairy tare (*Vicia hirsuta*), just for their leaves, which have multiple pairs of slender leaflets and branched tendrils (see p37). They add variety to a wild flower composition. As the plants wilt very quickly and are then difficult to handle, it helps, when gathered away from home, to place the material between the pages of a book in the position you want to use it and press it in the book. The best method is from the garden straight into the press. I find it worthwhile to consider these delicate plants, even though success is not always guaranteed.

This early spring growth as well as some bedstraws (*Galium*) and other small summer plants all have their uses (see p37). If not already present as so-called weeds, their introduction into the garden should be welcomed.

Bird's-foot Trefoil *Lotus corniculatus* (1:1)

Clover   In addition to the common white clover (*Trifolium repens*), I introduced the multi-leafed cultivar 'Purpurascens Quadrifolium' into my garden. It has a mixture of dark green leaves with three, four, five and six leaflets. Because of its symbolic "good luck" popularity, I use the four-leaf clover mainly on keyrings and greeting cards.

In wild flower compositions, white clover leaves become very useful without displaying the flowers as well. But if you wish to include the heads of dense small flowers, you have to choose carefully. If any of the lower flowers turn downwards and show signs of brown, the flowerhead has been in bloom for some time and is past its best. One side of a chosen head needs to be carefully thinned out to achieve the desired flatness in the press. Smaller, greenish and therefore not fully opened heads can be pressed whole.

Three-, four-, five- and six-leaf clover (1:1)

White Clover *Trifolium repens* 'Purpurascens Quadrifolium' (1:1)

Example of how a wild flower picture evolves

Stages of a composition with mainly wild carrot, cornflower, common poppy, meadow buttercup and hedgerow cranesbill

Astrantia The attractive garden cultivars of this wild flower present some challenges. A whole stem with branching stalks has to be taken apart, pressed and reassembled again in the right way. Flowerheads (made up of umbels of small flowers) surrounded by up to 20 petal-like bracts that give the flower its distinctive appearance, need to be detached from their stalks when pressed full-face. Smaller flowers can be pressed in profile with stalks attached. All stem-hugging small leaves need to be detached and pressed singly. All stems are too robust to be pressed whole, so they need to be carefully reduced along their full length (see p18). You have to decide at this point which side of the stalk you wish to show as the front when reassembling the plant, because this determines its final appearance. Include some not too large basal leaves as well.

Astrantia (cultivars range in colour from reds through pinks to white)

Fennel *Foeniculum vulgare*

Fennel This herb belongs to the carrot family, but unlike the umbels of the wild carrot, which can be pressed whole and full-face or halved in profile, the fennel umbel has to be taken apart. Its compound smaller umbels of tiny yellow flowers on short stalks need to be detached and pressed individually. The main umbel then has to be reconstructed to give its distinctive appearance. In order to achieve this, you need to press the stem with the umbel base intact, but in common with many other plants, the stem needs to be reduced from the back or split in half (see p18). Stalks of smaller umbels still in bud can be pressed whole for additional interest. Press also a selection of the more slender feathery leaves rather than the large mature ones.

***Kerria japonica* 'Pleniflora'** This double-flowered cultivar, one of several plants known as bachelor's buttons, is a valued shrub in gardens for its bright yellow blooms throughout spring. It is an easy candidate for the press. Flowers can be pressed full-face and in profile. Because the shrub carries blooms over a relatively long period, you can be selective and gather them when in prime condition. Once past their best, they are easy to recognize because petal tips and whole petals amongst the yellow start to turn white. Stems and leaves are pressed separately. Exceptions are the tips of new growth, where it is advantageous to press several leaves together as a unit.

Kerria japonica 'Pleniflora'

Elder *Sambucus nigra*

Elder The elder is a large shrub mainly growing wild in woodland edges and hedgerows. It has a variety of culinary uses, but it is not an obvious candidate for a pressed flower composition. Nevertheless, an experiment with a cluster of white elderflowers had a pleasing result. In order to make a representative picture, a variety of suitable, not too large leaves need to be considered. Some stem reduction will have to be made (see p18). And in addition to fully opened clusters of flowers, some still in bud will round off a picture. More about shrubs and trees under the heading Foliage (p60).

Grassy Areas

In a composition of wild flowers, I often include flowering grasses; they set the scene without becoming a dominant part. And because the finished work should ideally be as flat as possible, only loosely arranged panicles with a delicate appearance are suitable. Several grasses in meadows, lawns and pastures fit the bill. Some of those I use are shown on p37. When applying glue to the panicles of these grasses, the spikelets tend to stick together or overlap. Once glued down they can and should be pushed straight-away into their natural loose-look position.

Young grass plants can be pressed whole, even with the roots when carefully washed clean of soil. It is most useful material to have. Mature grass blades, indispensable in combination with flowering grass stems, are amongst the most difficult items to place into a press. Even the freshest, from the garden straight into the press, do not behave themselves, but bend or curl in all directions. As not only straight but also the curved or even folded over blades are of interest for a natural appearance, you have to put up with the fact that just a few blades can demand the whole space of a sheet in the press. Luckily, with grasses the pressing/drying process is short.

Creeping Soft Grass
Holcus mollis

Pressed blades of grass are good additions to depict flowers in their natural habitat, such as a lawn or meadow

Salad Burnet *Sanguisorba minor*
(complemented with clover and grass)

1–3) Meadow Grasses
Poa trivialis, P. annua, P. pratensis

4) Perennial Ryegrass
Lolium perenne

5) Red Fescue *Festuca rubra*

6) Yorkshire Fog *Holcus lanatus*

7) Greater Quaking Grass
Briza maxima

8) Velvet Bent *Agrostis canina*

9) Creeping Soft Grass
Holcus mollis

10) Common Bent
Agrostis capillaris

11) Field Woodrush
Luzula campestris

12) Sweet Vernal Grass
Anthoxanthum odoratum

In addition to grasses, here are some examples of useful plants for wild flower pictures gathered from grassy places and gaps in paved garden areas:

a) Herb Robert *Geranium robertianum*

b) Hedgerow Cranesbill
Geranium pyrenaicum

c) Lesser Trefoil *Trifolium dubium*

d) Cow Parsley
Anthriscus sylvestris 'Ravenswing'

e) Cleavers *Galium aparine*

f) Lady's Bedstraw *Galium verum*

g) Hairy Tare *Vicia hirsuta*

h) Rough-stalked Feather-moss
Brachythecium rutabulum

i) Broad-leaved Dock *Rumex obtusifolius*

Winter and Spring Flowers

Anemone   In profile, *Anemone blanda* can be pressed whole; but full-face, the flower and the stem with its leaves are pressed separately. With *Anemone* 'de Caen', all flowers are used full-face and the whorl of leaves is pulled off the tall stem. All parts are pressed individually and later put together again. Also some basal leaves are added. For the autumn-flowering *Anemone hupehensis* see p56.

Shorten the dome of pistils
before pressing *A.* 'de Caen'

Anemone 'de Caen'

Balkan Anemone
Anemone blanda (1:1)

Crocus The earliest small crocuses, like the snow crocus (*Crocus chrysanthus*), generally press well. Yellow ones hold their colour best. With blues and purples you have to expect some rejects. Even a white cultivar is worth pressing, because it prominently shows the anthers and saffron-coloured stigmas. Crocuses can be displayed in single or mixed colour groups. The flowers, although nearly transparent, accept glue quite well when applied with brush strokes from base to petal tips. Another reason why early crocuses are best is the fact that the blade-like leaves hold their colour. The leaves and some of the papery basal spathes are pressed singly and used in pictures sparingly. The earlier mentioned procedure (p23) of cutting away some overlapping material at the base of the design becomes relevant here.

The spring crocus (*Crocus vernus*) cultivars are a different matter. These later flowering large crocuses in purple, white, and striped, I find are not suitable as pressed flowers. The purple ones rarely hold their colour and the large leaves of these varieties turn brown in the flower press. Care has therefore to be taken, to press the right leaves (of the small crocus varieties) at the right time, otherwise you will end up with flowers and no leaves to match.

The crocus picture shown here is finished with grass and cleavers in order to round off the composition (see pp36–37).

Pressed material and stages of a snow crocus picture

Daffodil   Although flowers of the smaller varieties can be pressed whole, best results are achieved if stems, spathes and flowers, cut off at the point where the stem and spathe meet, are pressed separately. The blooms are pressed in profile if they have a trumpet-shaped corona, or full-face if they have a small cup-shaped one (see p87). In the latter case the tube behind the tepals (petals and sepals) is cut away, pressed separately, and if required added again later.

From larger daffodils, part of the ovary capsule is cut away and one outer tepal (sepal) peeled off from the same side. This treatment determines whether later the flower can be placed facing to the left or to the right. The exposed sticky part is then dabbed onto a kitchen towel before pressing. This is to prevent the flower sticking to the card in the press and getting damaged when trying to lift it off. Also, the papery spathe is peeled away and the stem split in half and dried off (see descriptions pp16–18). All these parts are pressed separately and reassembled in your presentation.

Unfortunately, daffodil leaves are not very suitable for pressing, turning pale, unless you apply the quite delicate process of shaving off the slightly raised midrib from the whole length of the underside (see diagram), so that the sap can be absorbed by the paper in the flower press. Alternatively, I sometimes use snowdrop leaves as substitutes for those of the daffodil (as depicted on p8).

A few grass blades, other low-growing early green plants like cleavers, and cow parsley, whose leaves appear as early as December give added interest. The dark green 'Ravenswing' is especially good. Cow parsley is also one of the earliest flowering plants of the parsley/carrot family.

Daffodil *Narcissus* and Cow Parsley
Anthriscus sylvestris 'Ravenswing'

Daffodil *Narcissus* and
Cleavers *Galium aparine*

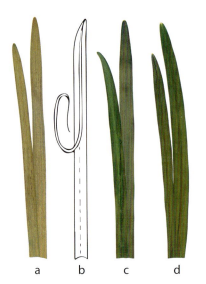

a) Daffodil leaf untreated

b) Daffodil midrib removal

c) Daffodil leaf after treatment

d) Snowdrop leaf untreated,
   used as a substitute

a   b   c   d

Bluebell The native English bluebell (*Hyacinthoides non-scripta*) is a protected species, but available in garden centres and from nurseries. The Spanish bluebell and cultivars are also present in gardens. Flowers and flower stems press well and the light blue colour of the bells turn to a deeper blue in the press. But here again, as with other bulbous spring flowers with leaves of similar linear leaf structure and high moisture content, pressing leaves leads to unpredictable results. The best choice for success is with the smaller and narrower leaves selected from the deepest green and pressed in a dry, warm air environment.

Spanish Bluebell
Hyacinthoides hispanica

Grape Hyacinth If the bell-shaped flowers from one side of the spike are carefully cut away, a grape hyacinth can be pressed with its stem. The dark blue flowers can bring contrast in combination with other spring flowers. Pressing the leaves needs some thinking ahead, i.e. gathering short ones well before the plants bloom, ideally during a cold spell, when your heating is on. Once the leaves become long and straggly, they turn paler as well and will not improve in the press.

Daffodil combined with
Grape Hyacinth

Grape Hyacinth
Muscari

Primula   The native primrose and most of its cultivars are suitable as pressed flowers, although in the wide colour range all reds turn into shades of brown. As with other flat-faced flowers with long tubes, it is best to pull the corolla out of the bell-shaped calyx, and to cut away the tube, before pressing full-face. In profile, flowers can be pressed with their tube. For all of these you will also need to press stalks with the calyx attached. The mix of buds, flowers full-face and in profile provide the material for a most natural effect. Leaves of various sizes from the basal rosette should also be pressed. The cowslip umbels look best when used whole in profile. Not included here is the *auricula* group, because the distinctively structured leathery leaves are not suitable for pressing.

Cowslip *Primula veris*

Primula elatior 'Gold Lace'

Primrose *Primula vulgaris*

Tulip It is worth experimenting with various cultivars. I found lily-flowered and multi-coloured tulips the most suitable. The tepals need to be pressed individually; three to four are enough for a cup-shaped flower display. Stamens and pistil are not included. The fleshy round stems need to be split into three sections, of which the two outer sections can be pressed; the inner section is discarded (see p18, diagram b).

Tulip *Tulipa*

Multi-headed Tulip

Lily-flowered Tulip

Cup-shaped Tulip

Leaves are a problem. The large basal leaves are not suitable and the smaller stem-based leaves have a mixed success rate of retaining a dark grey-green. It is a good idea to press smaller leaves from various tulips. I had the best results from leaves that are velvety, or those that have white edges; shiny ones tend to go grey. Smaller leaves from shop-bought tulips in winter turn out well in the quick pressing process. The tepals of all kinds of tulips are worth pressing for non-floral art (see pp72–81 and the colour chart p71).

Wallflower
Erysimum cheiri

Wallflower *Erysimum* 'Bowles's Mauve' (1:1)

Wallflower   The biennial wallflower is often planted in the garden for a display in spring, with flowers varying from cream, golden yellow to orange-brown and purple. This good colour range makes it a valuable plant for pressed floral art. All parts are pressed separately and can be arranged as shown. The dark green leaves from the purple varieties are best, because the leaves of the most common yellow-flowering plants fade too much. You can see this already on the growing plant.

An evergreen bushy perennial wallflower is *Erysimum* 'Bowles's Mauve'; its deep mauve flowers are smaller than the more usual garden varieties, but well worth having.

Lesser Celandine This is a wild flower that is mostly found in abundance under trees in gardens, flowering before the trees come into leaf. Flowers, which only open fully in sunshine, and leaves press well. I sometimes take off the yellow petals of a pressed flower and use the whorl of stamens, backed by the three sepals, as centres for other flowers whose stamens cannot be pressed (e.g. rose). The petals turn creamy-white within a year in storage. Because the small petals are not only strong in texture, but also have good opacity, they are excellent for the creation of bird pictures from plant materials (pp72–75).

The nearly opaque creamy-white petals
are indispensable in non-floral art

Lesser Celandine *Ficaria verna* (1:1)
(formerly *Ranunculus ficaria*)

Hellebore The flowers, stems (split), stemless leaves, and the larger, serrated basal leaves are all pressed separately. The flower is best used just after opening. Placed face down on the sheet, the flower needs to be pushed down at its centre with a finger: this opens it fully and positions the whorl of stamens evenly. Before pressing mature flowers face down, the seed capsule in the centre needs to be cut away carefully. This gives you material to create profile images. Before pressing a bud, most of its inner content has to be removed with tweezers. Caution: toxic – wear disposable gloves to handle these plants (see p24).

Hellebore *Helleborus* (1:1)

Hellebore *Helleborus*

46

Hairy Bittercress
Cardamine hirsuta

Rosettes The ground-hugging winter rosettes of hairy bittercress, shepherd's purse, dandelion and herb robert are very delicate to handle but they make nice displays. You may need to rinse off any soil and dry them carefully before placing in the press. Gaps resulting from leaves that have detached themselves can easily be closed again by glueing them back into place on the underside of the rosette.

Like the dandelion (p26), there are great variations in the leaves of the shepherd's purse – two rosettes are shown here side-by-side for comparison.

Herb Robert
Geranium robertianum

Shepherd's Purse *Capsella bursa-pastoris*

Aquilegia From late spring to early summer, this plant flowers in many varied shapes and colours. It is worth experimenting with, as all flowers retain good colour, but not all shapes are suitable for pressing. The foliage has a long season, from spring green to autumn yellow and purple. In common with many other plants, stem leaves have a different shape from basal ones, so you will need to press both types.

Aquilegia

Summer and Autumn Flowers

Pansy These are plants with all year round interest, and as the flowers of the many varieties can all be pressed equally well, they can often be found in books about pressed flowers. My recommendation is that you press not only fully opened flowers face on, but also buds in various stages of opening, stems with their green sepals just after petal fall and before the seed capsule is developed, and leaves in their various shapes. Keep an eye on the leaves, because not all varieties retain their deep green colour in the pressing process. After a while you will know beforehand which leaves might fail. With all parts of the plant at hand, displaying them is easy, either with flowers in a single colour or in a mixed colour scheme (see p86).

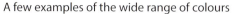
A few examples of the wide range of colours

Pansy
Viola × wittrockiana (1:1)

Passion Flower This intricately formed flower has interested me for a long time. Not least because it is such an unlikely candidate for the flower press. After a number of experiments, I found that I could actually retain most of the characteristics that gave the flower its name, i.e. the distinctive multi-coloured corona, which represents the crown of thorns, the three stigmas with styles representing three nails and the five anthers as symbols of the five wounds of Jesus on the cross. In order to achieve an acceptably flat flower, all I had to do was press the parts separately and re-assemble them later, i.e. the flower disc of the ten petals (actually five petals and five sepals of near equal shape, but different texture) that surround the corona, the stigmas and styles, and the stamens.

The goblet-shaped ovary and the fleshy nectary at the centre of the corona has to be cut away. In order not to damage the corona, a cut has to be made from behind the sepals to prise out the part where the stalk of the ovary meets the flower stem (see diagram). See also the section about nectar (p16); it is important for preparing a passion flower for pressing.

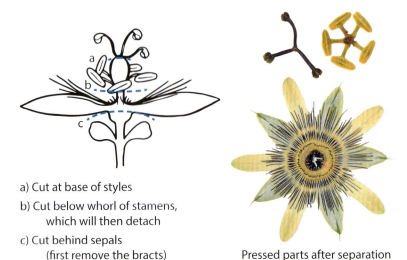

a) Cut at base of styles

b) Cut below whorl of stamens, which will then detach

c) Cut behind sepals (first remove the bracts)

Pressed parts after separation

Blue Passion Flower *Passiflora caerulea* (1:1)

50

For a good passion flower display, a selection of the following additional parts of the plant are needed: the five to seven-lobed leaves, the small half-moon shaped stipules from the base of the leaf stalks, sections of soft stem carefully split in half, and a few tendrils in their many shapes from curved to coiled. Mature tendrils have a wiry toughness; young ones are therefore more suitable for pressing. For a picture, I found it best to arrange and glue into place the stems and leaves first, leaving sufficient space to place the flower. I add the tendrils last for best all-round effect.

The above description refers to the most common, fairly hardy, blue passion flower (*Passiflora caerulea)*, but following the same principles of preparation for the press, other greenhouse or conservatory grown varieties can also be used.

Blue Passion Flower *Passiflora caerulea*

Passion Flower *Passiflora*
(greenhouse/conservatory variety)

St John's Wort The *Hypericum* genus of plants has many species and cultivars. The ground-covering rose of Sharon (*Hypericum calycinum*) and the taller shrub St John's wort 'Rowallane' are amongst the most common in gardens. The yellow flowers of both are very similar, except that the rose of Sharon has longer filaments, which can be difficult to handle. I therefore prefer 'Rowallane'. The flowers need to be taken apart and only the petals and whorl of stamens are pressed. Note the two-tone colour of the petals when assembling them again. Young leaves tend to turn brown in the press; it is therefore advisable to go for mature leaves in late summer or autumn. Stem reduction is necessary as described p18.

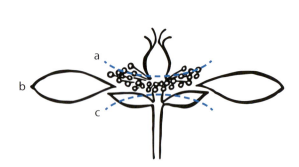

a) Cut above whorl of stamens
b) Petals detached
c) Cut between whorl of stamens and sepals

St John's Wort *Hypericum* 'Rowallane' (1:1)

Hydrangea The mopheads of the garden hydrangea can be arranged in two-dimensional form. Out of the many varieties, here I depict a pink one. The fleshy flowers with their four showy sepals are, however, very temperamental performers in the flower press. Each of the many flowers in one cluster needs to be pressed singly and reassembled later. For reasons I have not fully determined, whole batches of flowers can turn an unsightly brown colour, but others emerge from the press just fine. I suspect these unpredictable results occur because fresh flowers have a high moisture content.

Hydrangea macrophylla (in late summer)

Hydrangea macrophylla (in autumn)

I now wait until late summer and into autumn, when the still perfectly fine flowers have gradually turned dryer and therefore more robust and papery. The areas of the sepals exposed to sunshine turn red and purple, whereas the covered parts turn pale green. These contrasting coloured flowers can be arranged into interesting designs. To complete a picture, you also need a few of the branched stalks of the flowerhead and some of the smaller leaves, whose green is often also tinged with russet and crimson in autumn. For a main stem you can press a sliced off length of just the outer layer from a woody stem (see p18).

Rose Whole flowers of even the smallest varieties do not press well. Because of their firm centre, an even pressure cannot be applied to the petals. With roses I therefore take the most artistic licence, but without compromising the natural effect. All petals are pressed singly and put together again to form an open full-face rose. Depending on its diameter, I use a whorl of stamens from lesser celandine, meadow buttercup or *Anemone hupehensis* in the centre. I should be surprised if someone will even notice this substitution.

All leaves are usable, but I have found the young, deep wine-red ones in their various stages of growth press exceptionally well and allow for an appealing composition. The new growth also provides soft stem material with soft thorns for a realistic effect. If stems and leaf stalks are too thick, they should be split. If soft thorns (technically prickles) are lost in the process, they can be pressed singly and re-attached in advantageous positions in the overall design. It is amazing how often small details can make a big difference.

Pressed material and finished rose composition

Rose *Rosa*

Japanese Anemone   The white flowers in this autumn-flowering group will turn grey in the press. Of the pink-flowering varieties, the five-petalled ones such as 'Hadspen Abundance' are better suited for pressing than the multi-petalled cultivars, and those in the deepest pink will give the best results. The flower needs to be taken apart. Each actually has three petals and two sepals, but the latter have a slightly deeper colour and more pointed shape, giving the flower an unbalanced appearance. Therefore, ensure that the petals and sepals are reassembled in the right order. The quite prominent dome of pistils, surrounded by a whorl of stamens, is carefully cut short as shown in the diagram for spring-flowering anemones, p38.

Pot Marigold   A good candidate for pressed flower presentations. If not already present in the garden, it can easily be introduced from seed in their various cultivated colours and forms. For handling, follow the advice given for corn marigold (see p29), but note that the individual outer ray florets of single and double varieties are very delicate. The appropriate glueing down procedure for such flowers mentioned before (see p22), is advisable.

Japanese Anemone
Anemone hupehensis
'Hadspen Abundance'

Pot Marigold (double-flowered cultivar)
Calendula officinalis

Larkspur   The larkspur varieties are a good example of "flowers from vase to flower press", as they can provide material over many days with blooms at various stages of opening and colouring. Rather than using the large show blooms however, I prefer the smaller varieties I can grow successfully in the garden. Not only do the flowers vary in size and colour, but also in the shape of the leaves. Care has to be taken when choosing leaves of the more delicate, feathery kind. Only the ones with the deepest green press well. Older leaves and paler ones emerge from the press in faded olive green, giving less opportunity for striking contrast in presentation. Petals of white varieties can be set off against a darker background.

Larkspur *Consolida ajacis*

Gentian and Edelweiss These symbols of the alpine countries are protected in the wild, but nurseries and garden centres offer a variety of these plants originating from cultivated stock. More about gentians is explained under Nectar, Pollen and Sap (p16). Also, the position of stems and leaves has to be carefully worked out and the advice about overlapping material applied (see p23). Edelweiss is best displayed on a dark background and not covered with self-adhesive film.

Showy Chinese Gentian
Gentiana sino-ornata (1:1)

Edelweiss
Leontopodium nivale (formerly *L. alpinum*) (1:1)

Cyclamen Successful pressing is possible when their flowering period coincides with the heating season. Leaves of various sizes, buds and flowers are all pressed with stems attached. Uniform dark green leaves give better results than the more distinctly patterned ones. It is important to start the pressing process early in the day by applying only gentle pressure and to increase this gradually, spread over a number of hours.

Cyclamen (1:1)

Foliage

Picking up some colourful autumn leaves in parks or nearby woods is something many people cannot resist. Unfortunately, the beauty only lasts a short while. Pressing autumn leaves is, with some exceptions, not very successful. In general, leaves are too large, too hard in texture, and most important of all, pressing does not stop the process of fading. For a display of mixed coloured leaves I found by chance that spring and early summer provides interesting material. For instance, horse-chestnut leaves that grow in spring on the trunks of mature trees are perfect miniature versions of the large leaves. They grow in different shades of green and sometimes there may be some with very little chlorophyll, which turn yellow or ochre in the press (see p19).

Young leaves from different species of oak provide a variety of interesting shapes and colours. In summer, new leaves on the English oak in hedgerows often have tinges of orange to deep red. Young leaves from the maple family of trees also vary in colour. It is interesting to experiment with all kinds of leaves.

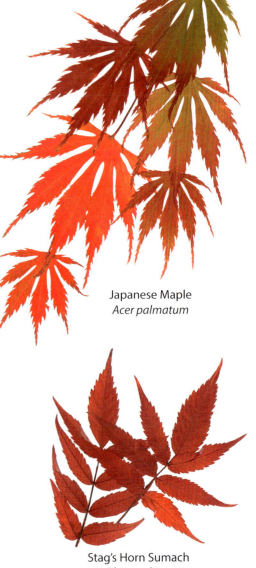

Japanese Maple
Acer palmatum

Spring leaves
imitating autumn colours

Pin, Turkey, and English Oak (young leaves)
Quercus palustris, Q. cerris, and Q. robur

Stag's Horn Sumach
Rhus typhina

Maple The leaves in combination with clusters of small flowers or young winged fruit merit a display on their own. Pressed leaves, especially of the young Norway maple, appear nearly transparent; they can often be so thin that they need careful handling. The Japanese maple varieties provide leaves from spring to autumn in green, yellow, orange, and from bright red to burgundy. Often the foliage of basal shoots on grafted ornamental shrubs and trees is different from the main plant, but worth considering on their own or for a composition of mixed leaves.

Norway Maple
Acer platanoides

Field Maple *Acer campestre*
(leaves and winged fruit)

Japanese Maple *Acer palmatum*
(leaves and winged fruit)

Japanese Maple *Acer palmatum*
(flowering)

Basal shoots from an *Acer
palmatum* cultivar

Robinia pseudoacacia

Smoke Tree In addition to the autumn colour-retaining maple leaves, the smoke tree leaves provide an array of colours.

Smoke Tree
Cotinus coggygria

Garden Strawberry
Fragaria × ananassa

The garden strawberry An interesting range of summer and autumn colours can be found not only on shrubs and trees, but also on plants closer to the ground.

Because strawberry leaves are much more prone to damage, you have to inspect them regularly and pick them at the right moment for use in the flower press.

Flowering ornamental shrubs and trees These should not be overlooked for their potential. I have already mentioned the difficulties with flowers from the pea family in general (p31), and these extend to large shrubs and trees. Unsurprisingly, attempts to press foliage and flowers of *Wisteria* have defeated me. The ornamental tree *Robinia,* however, can provide quite interesting foliage and the clusters of white flowers, which turn a cream colour in the press, present a good and lasting feature.

Bloody Cranesbill
(cultivar)
Geranium sanguineum

Meadow Cranesbill (cultivar)
Geranium pratense (above and right)

Cranesbill Excellent examples of colour-changing foliage can be found amongst its various wild forms and cultivars.

Bloody Cranesbill
Geranium sanguineum

European Larch I also experimented with conifers, and was only successful with the deciduous european larch. Because the soft needles grow in clusters along thin twigs, these clusters can be detached individually and the thickness at their base carefully reduced on one side. Treated in this way, they become pressable. From the twigs, too, the bark can be cut away on one side, making it possible to peel off the rest of the bark from the wooden centre. Pressed, a length of bark is flattened, but retains the appearance of a twig. The clusters of pressed needles are then placed along the strips of bark in a way that resembles a natural look as closely as possible. Such arrangements become most useful in pictures of birds associated with the environment of larch trees (see pp74-75).

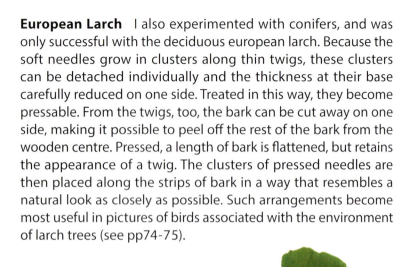

Ginkgo (Maidenhair Tree)

The *Ginkgo biloba* is the only living tree of its kind that has survived from over 200 million years ago when dinosaurs roamed the earth. First introduced to Europe in the 18th century from China via Japan as an ornamental tree, it is now available from nurseries to everyone. For a natural pressed display of the fan-shaped leaves, the preparations are very similar to those described for the larch.

Ginkgo (Maidenhair Tree)
Ginkgo biloba

European Larch
Larix decidua

Wild Vines From vines, such as virginia creeper, young green shoots can be pressed all summer. They also retain their autumn colours, from yellow to deep purple-red, after pressing. Here, too, are opportunities for experiments and new discoveries.

In a tone-on-tone display, the features of these climbers are interesting in themselves, but can be enhanced by deliberately showing summer and autumn foliage together.

Virginia Creeper
Parthenocissus quinquefolia (1:1)

Virginia Creeper
(flowering)

Unusual Experiments

After working with a range of familiar flower material for a while, you get quite a good idea of what is suitable and what is not. In between lies all the plant material worth experimenting with. Here are three examples, with more one-off experiments like mossy saxifrage, poinsettia and broom included in the Gallery (pp82–87).

Chrysanthemum 'Anastasia Green'

Having received some of these flowers in a birthday bouquet, I was enticed to try one in the flower press. For a two-dimensional display I did not need all of the countless individual ray florets, but a sufficient quantity of various sized ones had to be pressed to gain a good stock to use later. The cup of scale-like bracts is handled in the same way as described for cornflowers and sunflowers (p28). The curved florets pressed well, although in various shades. This turned out to be an advantage in the reconstruction of the flowerhead in profile. With the fresh flower firmly in mind (if in doubt refer to a photo), the creation of a representative image took patience and time. After all, a considerable number of florets needed to be placed individually in order to create an acceptable whole.

Chrysanthemum 'Anastasia Green' (1:1)

Wild Teasel   A fully developed, tall-growing teasel is far from suitable for pressing. Nevertheless, it is possible to depict the characteristics of the plant by choosing some still underdeveloped teasel stalks growing off the main stem lower down. All the already mentioned handling techniques with other plants are used, i.e. the extent of taking material apart, stem reduction (p18), etc. The long spiky bracts need to be detached first. Then you have to experiment how far the oval-shaped flowerheads can be reduced without falling apart. After pressing all parts individually, you can choose how much to include for a natural appearance.

Peony   Another experiment I made was with the large multi-petalled flower of the peony. It was fortunate that I had pressed all the many petals of the flower. I have no explanation why a large number of petals lost their colour completely or partially, whereas others retained their colour in various shades perfectly. Anyway, I had enough material to create several pictures. To add a bud, I carefully took one apart and pressed the outer layers singly, creating later the image of a bud with as many or as few round sepals as necessary.

Peony *Paeonia*

Wild Teasel *Dipsacus fullonum*

Common Poppy
Papaver rhoeas

Conservation of Wild Plants

At a time when an increasing number of wild plants come close to extinction and need our protection, it is important to make sure that only the most common plants are considered for the purpose of pressing, even when only parts of a plant are used and carefully gathered, i.e. without damaging the root system and/or depriving the plant of seed production.

All plants mentioned in this book are either present in abundance in the wild or can be successfully grown in wild flower beds in the garden or even in containers. E.g., the cornflower and poppy are under the heading Wild Flowers (p25), referring primarily to their origin. But now, as they have nearly disappeared from the fields, conscious efforts need to be made to grow them from seed in a garden environment.

On the other hand, there are wild flowers, like the dandelion, stinging nettle, etc. that are treated in the garden as unwelcome and are routinely eradicated. Using these so-called weeds in the pressed plant craft does no harm to the wild flora in any way. By gathering a few seeds of the wild carrot from the roadside, I introduced this plant to my garden. It has many uses in pressed flower applications.

As for leaves from shrubs and trees, my suggestions tend to favour the young leaves from the short-lived basal growth near tree trunks or lower parts of older trunks, and also shrubs and hedges that are regularly pruned or trimmed anyway.

And as for your garden plants, I am sure that you will take good care of these and harvest material in the same way as cutting flowers for the vase. It is a good idea to introduce plants into the garden with the flower press in mind, and also to accept contributions from helpful friends and neighbours.

It is also quite possible to grow many plants from seed in pots, even trees such as horse-chestnut, oak, maple, etc. Our family has several that are decades old. In pots or containers they will remain small and provide leaves every year.

The *BSBI Code of Conduct* is available from the Botanical Society of Britain and Ireland. Other countries should have their own information about picking, collecting, photographing and enjoying wild plants, and their protection under the law.

I presume that anyone who picks up this book has already an interest in the natural world that sustains us all. May it deepen the appreciation of the beauty and riches in every green space we have or create.

Taking Pressed Designs Further

A plant's black silhouette on a white or pastel-coloured background can be quite impressive, and give interesting shapes to work with. You can turn a digital scan or photo into a silhouette, or create other effects, using image-editing tools.

Hairy Bittercress
Cardamine hirsuta
(silhouette effect)

Stinging Nettle *Urtica dioica*
(composition with silhouette effect)

Virginia Creeper
Parthenocissus quinquefolia
(original and three examples of digital image-editing)

Personal Stationery

As well as creating original greeting cards from pressed plant material, you could use a design for other purposes by making a digital image of it.

With a colour or mono printer you could print your own letter paper or make notelets (e.g. A4 sheet of 100 g/m² paper folded in half lengthwise, then folded to A6). Or, use your designs online …

If you intend to frame or cover your designs, it is best to scan or photograph them before doing so. Scanning gives a more even light across the design, and is another reason to glue down all material as recommended. Even then, check your scan and scanner for any tiny bits that may have fallen off.

Thank you

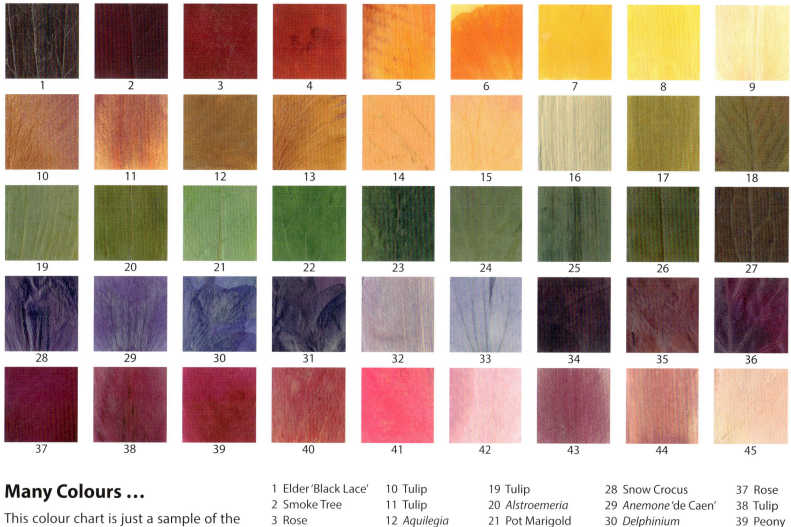

Many Colours …

This colour chart is just a sample of the great palette nature provides for us via the flower press. It emphasizes the point that neither dyed nor painted flowers need to be used in this craft.

1 Elder 'Black Lace'	10 Tulip	19 Tulip	28 Snow Crocus	37 Rose
2 Smoke Tree	11 Tulip	20 *Alstroemeria*	29 *Anemone* 'de Caen'	38 Tulip
3 Rose	12 *Aquilegia*	21 Pot Marigold	30 *Delphinium*	39 Peony
4 Tulip	13 Tulip	22 *Aquilegia*	31 *Aquilegia*	40 Tulip
5 Tulip	14 Cherry	23 Stinging Nettle	32 Tulip	41 Rose
6 Daffodil	15 Rose	24 *Aquilegia*	33 Pansy	42 Peony
7 Daffodil	16 Tulip	25 Tulip	34 Pansy	43 Tulip
8 Daffodil	17 *Alstroemeria*	26 Holly	35 Pansy	44 Tulip
9 Tulip	18 Strawberry	27 *Clematis*	36 *Heuchera*	45 Poppy

Creating Non-floral Pictures

Over time, the accumulation of surplus pressed material led to the idea to use it for non-floral art. I found that its qualities make it an excellent medium for creating a picture. Bits of plants can do much that would be more difficult to achieve otherwise.

Depicting Birds Sketch your chosen bird either free-hand or trace the outlines from a picture and start pasting. Here, too, I recommend the use of liquid paper glue (see p10). The shaping of material is done with a craft knife. An important aspect is the right sequence of overlapping parts. Observe the placement of a bird's feathers. Start at the tail and work towards the body and head. Usually, the head is best shaped from the back and neck towards the beak.

I found white shades in plant materials to be in short supply. Therefore, it takes a little effort to find and successfully press white flower petals that actually stay white, such as those of white-flowering larkspur. And, for example, the yellow petals of the lesser celandine that turn white (see p45) are most useful, as are the petals of pale blue or white love-in-a-mist (*Nigella damascena*). They have a feathery appearance and after pressing turn white/grey within a few months.

For a picture of this kind, the whole setting has to be thought of as well. You may need to choose and press foliage, twigs, bark, etc. for this specific purpose.

Nuthatch in Ivy
(complemented by a thin piece of bark)

Wilted and discoloured tulip leaves
make very good substitutes for wood (1:1)

Robin on Hawthorn

Long-tailed Tits
on Field Maple

Blue Tit on Birch

Great Spotted Woodpeckers
(complemented by
a thin strip of bark)

Goldcrests on Larch

Owl in Landscape
(fantasy composition)

Sparrow on Hornbeam

Crested Tit on Larch

Sample collection of stock material

Creating landscapes Here everything is left to the imagination. With naturally curved stalks of leaves and flowers, you can create contours in the landscape. I use predominately pressed clover stalks for this purpose. Large pastel-coloured petals can suggest hills and valleys, fields, lakes and skylines. Various types of leaves and branched stalks can represent trees and shrubs. Once you have worked with pressed plants for a while and have accumulated a good stock of material, you will find plenty to feed your creativity. As the outlines are made through the natural shape of plant material and by shaping with the craft knife, not a single pencil line is needed in these pictures.

Spring Landscape (1:1)

Winter Landscape (1:1)

Storm Approaching (1:1)

Widening the range   Some examples of how pressed plant material can become useful in diverse art projects.

Cherry blossom created from small rose petals
and dark flat twig sections (1:1)

A few iris leaves combined with clusters of emerging winged seeds of the ash tree and a rose petal is all you need to create an oriental image (right). It takes a lot of practice and the right tools to create a picture in the style of an ancient Chinese brush painting. But with pressed plant material you can, without any painting skills, imitate aspects of this distinctive art.

Bamboo imitation using young
sprays of winged ash seeds
and iris leaves

Sunset (1:1)

The Kite (1:1)

My Cottage (1:1)

Rainbow Trout
(marquetry imitation, see wood effects p72)

Frog in Pond Margin (1:1)

Gallery

The majority of my pressed plants have over the years been used for the creation of greeting cards as a means of raising funds for charity. Most of the original compositions shown in this book are therefore designs fitting into a limited predetermined space. I mainly use smooth or slightly textured single fold white cards, with or without printed borders.

The following examples show how diverse material can be arranged.

Daisy *Bellis perennis* 'Flore Pleno' and Forget-me-not *Myosotis*

Trailing Lobelia
Lobelia erinus

Cranesbill (pink)
Geranium

Pelargonium

Cupid's Dart
Catananche caerulea

Mossy Saxifrage
Saxifraga × *arendsii*

Orange Hawkweed (see also p27)
Pilosella aurantiaca

Larkspur
Consolida ajacis

Cosmos
Cosmos bipinnatus

Prairie Mallow
Sidalcea hybrida

White Dead-nettle
Lamium album

Summer Meadow
(Bird's-foot Trefoil and White Clover)

Verbena (annual)
Verbena × hybrida

Common Sunflower
Helianthus annuus

Pasque Flower
Pulsatilla vulgaris

Salad Burnet
Sanguisorba minor

Scaly Male Fern
Dryopteris affinis

Maidenhair Fern
Adiantum raddianum

Collection of various leaves

Japanese Maple
Acer palmatum

Geranium wlassovianum

Hellebore
Helleborus

Snowdrop
Galanthus

Snow Crocus
Crocus chrysanthus

Tulip
Tulipa

Primula (garden cultivar)

Striped Squill
Puschkinia scilloides

Alum Root
Heuchera

Pansy
Viola × wittrockiana

Poinsettia
Euphorbia pulcherrima

Broom
Genista

Hairy Bittercress
Cardamine hirsuta

Hedgerow Cranesbill
Geranium pyrenaicum

Elder
Sambucus nigra 'Black Lace'

Feverfew
Tanacetum parthenium

Daffodil (multi-headed)
Narcissus

Love-in-a-mist
Nigella damascena

Daffodil
Narcissus 'Rip van Winkle'

Glossary

Anther   Part of the stamen that bears the pollen.

Basal   Emerging from ground level, i.e. the base of a plant.

Bract   Small, leaf-like plant part just below the flowers, sometimes numerous and overlapping (e.g. cornflower).

Calyx   The sepals as a whole, and term used when the sepals are fused together to form a tube or cup (e.g. primrose).

Chlorophyll   Green pigment found in plants.

Composite   Flowerhead made up of many small florets (e.g. daisy, sunflower).

Corolla tube   Petals joined together in a tubular shape.

Corona   Trumpet or cup-shaped structure (e.g. daffodil) or radial appendages between petals and stamens (e.g. passion flower).

Cultivar   A cultivated plant selectively propagated to enhance certain characteristics.

Dioecious   Having the male and female flowers on separate plants (e.g. stinging nettle).

Filament   The stalk of the stamen that supports the anther.

Floret   Small flowers in the head of a composite flower, i.e. disc florets surrounded by ray florets, all ray florets or just disc florets (see diagram p28). Small flowers making up the spikelets in grasses.

Grafting   A type of vegetative propagation that joins together different plants for certain characteristics.

Midrib   Central vein providing water and nutrients to a leaf.

Nectar   Sweet substance secreted by many flowers to attract insects. The place where a flower stores its nectar is called the nectary (prominent in a passion flower).

Ovary   Female reproductive part of a flower that develops into a fruit containing a seed or seeds.

Panicle   Loose flower formation with flowers on stalks branching out from a main stem.

Petal   Petals are parts of a flower (collectively known as corolla) that surround the reproductive organs.

Pistil   Collective name for stigma, style and ovary.

Pollen   Tiny grains containing male cells produced in the anthers.

Prickle	Sharp-pointed structure growing from a stem or leaf (e.g. rose).
Receptacle	Tip of a flower stalk/stem to which the organs of the flower are attached (cup-shaped in e.g. cornflower, sunflower).
Rosette	A circular structure of leaves at the base of a plant.
Sap	Fluid transporting water and nutrients throughout a plant.
Sepal	Sepals (petal-like and usually green) form a ring/whorl (calyx) just below the petals, acting as protection for the bud and support when open. If near equal in appearance to petals, they are collectively called tepals (e.g. daffodil, tulip).
Spathe	Leaf-like bract enclosing a flower in bud (e.g. crocus, daffodil).
Spike	A formation of flowers without stalks surrounding a main stem. Sometimes with short stalks (e.g. grape hyacinth).
Spikelet	Several florets branching off a flower-spike (e.g. grasses).
Stalk	Subsidiary stem branching off a main stem.
Stamen	The male reproductive part of a flower, including the filament and anther.
Stem	Main structure of a plant bearing leaves, a flower or arrangement of flowers.
Stigma	The uppermost part of the female organ that receives the pollen.
Stipule	Scale or leaf-like growth at the base of leaf stalks where they meet the stem.
Style	Elongated part of the pistil connecting the stigma and ovary.
Tendril	A modified stem or leaf that enables a plant to climb.
Tepal	Collectively the petals and sepals when they look alike (e.g. daffodil, tulip).
Twig	A small thin branch of a woody plant such as a tree or shrub.
Umbel	Cluster of flowers on short stalks radiating from a single point on a stem (e.g. wild carrot, fennel, cowslip).
Whorl	Circular arrangement of organs around one point of a stem or flower (e.g. leaves, stamens).

Bibliography

Axon, John (ed.), *Country Crafts* (London: Macdonald Educational Ltd, 1979)

Backhouse, Janet, *The Illuminated Page: Ten Centuries of Manuscript Painting in the British Library* (London: British Library, 1997)

Blamey, Marjorie, and R. Fitter and A. Fitter, *Wild Flowers of Britain and Ireland* (second edition, London: Bloomsbury Publishing, 2013)

Blossfeldt, Karl, *Art Forms in Nature: The Complete Edition* (Munich: Schirmer/Mosel GmbH, 1999)

Bologna, Giulia, *Illuminated Manuscripts: The Book Before Gutenberg* (London: Thames and Hudson, 1988)

Brickell, Christopher (ed.), *RHS Gardeners' Encyclopedia of Plants & Flowers* (London: Dorling Kindersley Ltd, 1990)

Buchheim, Diethild, and Lothar-Günther Buchheim, *Blätter Menagerie: Brunnen-Reihe 132* (Freiburg: Christophorus-Verlag, 1978)

Christiansen, M. Skytte, *Grasses, Sedges and Rushes in Colour* (Poole: Blandford Press, 1979)

Eno, David, *Pressed Flowers* (Winchester: Juniper Press, 1980)

Evans, Jane, *Chinese Brush Painting: A Complete Course in Traditional and Modern Techniques* (London: William Collins, 1989)

Fenton, Joyce, *Pressed Flower Craft* (Tunbridge Wells: Midas Books, 1980)

Fitter, Richard, and Alastair Fitter (text) and Ann Farrer (illust.), *Collins Guide to the Grasses, Sedges, Rushes and Ferns of Britain and Northern Europe* (London: Collins, 1984)

Flowers, Diane, *Preserving Flowers: Dried & Pressed Floral Designs for Every Season* (New York/London: Sterling Publishing, 2006)

Harrap, Simon, *Harrap's Wild Flowers* (London: Bloomsbury Publishing, 2013)

Innes, Miranda, and Clay Perry, *Medieval Flowers* (London: Kyle Cathie Ltd, 1997)

King, Christabel, *The Kew Book of Botanical Illustration* (Tunbridge Wells: Search Press Ltd, 2015)

Koehler, Horst, *Das bunte Blumenbuch* (Gütersloh: C. Bertelsmann Verlag, 1958)

Lee, Lawrence, et al., *Stained Glass* (London: Mitchell Beazley Publishers Ltd, 1976)

Martin, W. Keble, *The Concise British Flora in Colour* (London: Ebury Press and Michael Joseph, 1974)

McDowall, Pamela, *Pressed Flower Pictures* (Guildford and London: Lutterworth Press, 1975)

Nicol, Kit, *Flowers for Pleasure: Craft and Decorative Ideas* (Guildford and London: Lutterworth Press, 1974)

Phillips, Roger, and Martyn Rix, *The Botanical Garden* (London: Macmillan Publishers Ltd, 2002)

Puckett, Sandy, *Fragile Beauty: The Victorian Art of Pressed Flowers* (New York: Warner Books Inc, 1992)

Reader's Digest Association, *Field Guide to the Wild Flowers of Britain* (London: 1981)

Reader's Digest Association, *New Encyclopedia of Garden Plants & Flowers* (London: 1997)

Robertson, Debora, *Gifts from the Garden* (London: Kyle Books, 2012)

Sheen, Joanna, *The Microwave Pressed Flower Manual* (London: Aurum Press, 1998)

Sherwood, Shirley, and Martyn Rix, *Treasures of Botanical Art* (Kew: Royal Botanic Gardens, 2008)

Šikula, Jaromír, and Vojtěch Štolfa (illust.), *Grasses*, trans. Olga Kuthanová (London: Hamlyn, 1978)

Smith, Margaret, *Pressed Flowers* (London: Search Press Ltd, 1976)

Spencer, Margaret, *Pressed Flower Decorations* (London: William Collins, 1975)

Spohn, Margot and Roland, *Black's Nature Guides: Wild Flowers of Britain and Europe* (London: A&C Black Publishers Ltd, 2008)

Spohn, Margot and Roland, *Kosmos-Naturführer: Was blüht denn da?* (Stuttgart: Franckh-Kosmos Verlags-GmbH & Co. KG, 2015)

Starý, František and Václav Jirásek, *Herbs: A Concise Guide in Colour*, trans. Olga Kuthanová (London: Hamlyn, 1973)

Streeter, David, *Collins Wild Flower Guide* (second edition, London: William Collins, 2016)

Watkins, Doris, *Blumen pressen* (Augsburg: Augustus Verlag, 1999)

Wilson, Ron, *The Hedgerow Book* (Newton Abbot: David & Charles, 1979)

Flora (London: Stanley Gibbons Magazines Ltd) magazine articles:
    Handley, Ivora, 'Pressed Flower Accessories', September–October 1983
    Hildebrandt, Bernd, 'Greeting Cards for all Occasions', November–December 1983
    Hildebrandt, Bernd, 'Pressed Flower and Plant Designs with a Difference', May–June 1980
    Hildebrandt, Bernd, 'Pressed Spring Leaves for Autumn Pictures', March–April 1980

Flora (Ringwood: Stanley Gibbons Magazines Ltd) magazine articles:
    Handley, Ivora, 'Two Pressed Flower Designs', September–October 1987
    Minns, Arline, 'Silver leaves', July–August 1987

The bird picture designs are based on and adapted from various images in books, calendars and cards.

Index

* Plants not illustrated

A

Acer campestre 61
  palmatum 60, 61, 85
  platanoides 61
Adiantum raddianum 84
Aesculus hippocastanum 19
Agrostemma githago* 29
Agrostis canina 37
  capillaris 37
Alstroemeria 13, 14, 71
Alum Root 86
Anemone 38
  blanda 38
  'de Caen' 38, 71
  hupehensis 38, 54, 56
    'Hadspen Abundance' 56
Anemone, Balkan 38
  Japanese 56
Anther 39, 50
Anthoxanthum odoratum 37
Anthriscus sylvestris 'Ravenswing' 7,
    37, 40
Aquilegia 48, 71
Ash, Common 15, 78
Astrantia 34
Auricula* 42

B

Bachelor's Buttons 35
Bark 18, 64, 72, 74
Basal 34, 38, 39, 42, 43, 46, 48, 61, 68

Bedstraw, 31
  Lady's 37
Bellflower* 14
Bellis perennis 'Flore Pleno' 82
Bent, Common 37
  Velvet 37
Birch 73
Bittercress, Hairy 20, 47, 69, 87
Blade (of grass) 23, 36, 40
Bloom 7, 11, 13, 14, 25, 35, 40, 57
Bluebell, English* 41
  Spanish 41
Brachythecium rutabulum 37
Bract 26, 28, 34, 50, 66, 67
Briza maxima 37
Broom 66, 86
Bud 7, 12, 26–28, 34, 35, 42, 46, 49,
    59, 67
Buttercup, Meadow 24, 33, 54

C

Calendula officinalis 56
Calyx 16, 26, 42
Campanula* 14
Campion, Red 25
Capsella bursa-pastoris 47
Cardamine hirsuta 47, 69, 87
Carrot, 34, 40
  Wild 11, 12, 33, 34, 68
Catananche caerulea 82
Celandine, Lesser 45, 54, 72
Centaurea cyanus* 28
  montana 28
Cherry* 71
Chlorophyll 60
Chrysanthemum 'Anastasia Green' 66

Cinquefoil, Creeping 10
Cleavers 37, 39, 40
Clematis 15, 71
Clover, 25, 32, 36, 76
  White 32, 83
    'Purpurascens Quadrifolium' 32
Composite 26, 28, 29
Consolida ajacis 57, 83
Corncockle* 29
Cornflower, 28, 33, 66, 68
  Perennial 28
Corolla tube 16, 42
Corona 16, 40, 50
Cosmos bipinnatus 83
Cotinus coggygria 62
Cowslip 42
Cranesbill, 63, 82
  Bloody 63
  Hedgerow 27, 33, 37, 87
  Meadow 63
Crocus, 20, 39
  Snow 39, 71, 85
  Spring* 39
Crocus chrysanthus 39, 85
  vernus* 39
Cupid's Dart 82
Cyclamen 59

D

Daffodil 8, 16, 18, 24, 40, 41, 71, 87
  'Rip van Winkle' 87
Daisy, 20, 25
  Crown 29
  'Flore Pleno' 82
Dandelion 18, 20, 26, 27, 47, 68
Daucus carota 11, 12

Dead-nettle, White 83
*Delphinium** 71
Dioecious 30
Dipsacus fullonum 67
Dock, Broad-leaved 25, 37
Dryopteris affinis 84

E
Edelweiss 58
Elder 35
 'Black Lace' 15, 71, 87
Erysimum 'Bowles's Mauve' 44
 cheiri 44
Eschscholzia californica 27
Euphorbia pulcherrima 86

F
Feather-moss, Rough-stalked 37
Fennel 34
Fern, Maidenhair 22, 84
 Scaly Male 84
Fescue, Red 37
Festuca rubra 37
Feverfew 87
Ficaria verna 45
Filament 52
Floret, 28, 66
 Disc 28
 Ray 28, 29, 56, 66
Flower 7, 8, 11–14, 16, 18, 20–46,
  48–54, 56, 57, 59, 61, 62,
  65–68, 71, 72, 76
Foeniculum vulgare 34
Foliage 14, 48, 60–63, 65, 72
Forget-me-not 82
Fox-and-cubs 27

Fragaria × *ananassa* 62
Fraxinus excelsior 15
Fruit, 7
 Winged 15, 61

G
Galanthus 85
Galium 31
 aparine 37, 40
 verum 37
Genista 86
Gentian, 16, 58
 Showy Chinese 58
Gentiana 16
 sino-ornata 58
Geranium 82
 pratense 63
 pyrenaicum 27, 37, 87
 robertianum 37, 47
 sanguineum 63
 wlassovianum 85
Ginkgo biloba 64
Glebionis coronaria 29
 segetum 29
Graft 61
Grape Hyacinth 8, 41
Grass, 23, 25, 36, 37, 39, 40
 Creeping Soft 36, 37
 Greater Quaking 37
 Meadow 25, 37
 Sweet Vernal 37

H
Hawkweed, 25, 27
 Orange 27, 83
Hawthorn 73

Helianthus annuus 84
 rigidus 28
 tuberosus 28
Hellebore 18, 24, 46, 85
Helleborus 46, 85
Herb Robert 37, 47
Heuchera 71, 86
Hieracium 27
 aurantiacum 27
Holcus lanatus 37
 mollis 36, 37
Holly* 71
Hornbeam 75
Horse-chestnut 19, 60, 68
Hyacinthoides hispanica 41
 *non-scripta** 41
Hydrangea 18, 53
 macrophylla 53
Hypericum 52
 *calycinum** 52
 'Rowallane' 52

I
Iris, 78
 Siberian 14
Iris sibirica 14
Ivy 72

J
Jerusalem Artichoke 28

K
Kerria japonica 'Pleniflora' 35

L
Laburnum 24

Lamium album 83
Larch, European 18, 64, 74, 75
Larix decidua 64
Larkspur 57, 72, 83
*Lathyrus latifolius** 31
  odoratus 14
  *pratensis** 31
Leaf/Leaves 7, 8, 11–15, 18, 20–23,
    26–32, 34, 35, 38–49, 51–54,
    57–62, 64, 68, 72, 76, 78, 85
Leontopodium alpinum 58
  nivale 58
Lily, Peruvian 13
Lobelia erinus 82
Lobelia, Trailing 82
Lolium perenne 37
Lotus corniculatus 31
Love-in-a-mist 72, 87
Lupin* 31
*Lupinus** 31
Luzula campestris 37

M
Maidenhair Tree 64
Mallow, Prairie 83
Maple, 7, 18, 60–62, 68
  Field 61, 73
  Japanese 60, 61, 85
  Norway 61
Marigold, Corn 29, 56
  Pot 56, 71
Midrib 18, 40
Muscari 8, 41
Mycelis muralis 15
Myosotis 82

N
Narcissus 8, 16, 40, 87
  'Rip van Winkle' 87
Nectar 16, 17, 50
Nettle, Stinging 24, 30, 68, 69, 71
Nigella damascena 72, 87

O
Oak, English 60, 68
  Pin 60
  Turkey 60
Ovary 27, 40, 50

P
Paeonia 67
Panicle 36
Pansy 7, 20, 22, 49, 71, 86
Papaver nudicaule 27
  rhoeas 27, 68
Parsley, Cow 40
  'Ravenswing' 7, 37, 40
Parthenocissus quinquefolia 65, 69
Pasque Flower 84
Passiflora 51
  caerulea 50, 51
Passion Flower, 16, 20, 50, 51
  Blue 50, 51
Pea, 31, 62
  Everlasting* 31
  Sweet 14
Pelargonium 82
Peony 18, 67, 71
Periwinkle, 16
  Greater 17

Petal 14, 16, 20–22, 27, 35, 39, 40, 45,
    50, 52, 54, 56, 67, 72, 76, 78
Pilosella aurantiaca 27, 83
Pistil 38, 43, 56
Poa annua 37
  pratensis 37
  trivialis 37
Poinsettia 66, 86
Pollen 16
Poppy, 20, 27, 68, 71
  California 27
  Common 27, 33, 68
  Iceland 27
Potentilla anserina 12
  reptans 10
Prickle 18, 54
Primrose 42
Primula 16, 42, 86
  elatior 'Gold Lace' 42
  veris 42
  vulgaris 42
Pulsatilla vulgaris 84
Puschkinia scilloides 86

Q
Quercus cerris 60
  palustris 60
  robur 60

R
Ranunculus acris 24
  ficaria 45
Receptacle 26, 28, 29
Rhus typhina 60
Robinia pseudoacacia 62

Root 7, 36, 68
Rosa 8, 55
Rose 8, 18, 45, 54, 55, 71, 78
Rose of Sharon* 52
Rosette 20, 42, 47
Rumex obtusifolius 25, 37
Ryegrass, Perennial 37

S

Salad Burnet 36, 84
Sambucus nigra 35
  'Black Lace' 15, 87
Sanguisorba minor 36, 84
Sap 16, 40
Saxifraga × arendsii 82
Saxifrage, Mossy 66, 82
Seed 24, 27, 29, 31, 56, 68, 78
  Capsule 46, 49
  Pod 31
Seedhead 7
Sepal 27, 40, 45, 49, 50, 52, 53, 56, 67
Shepherd's Purse 20, 25, 47
Sidalcea hybrida 83
Silene dioica 25
Silverweed 12
Smoke Tree 62, 71
Snowdrop 8, 40, 85
Spathe 39, 40
Spike 41
Spikelet 36
Squill, Striped 86
Stalk 7, 12, 13, 18, 34, 42, 50, 51, 53, 54, 67, 76
Stamen 8, 16, 43, 45, 46, 50, 52, 54, 56

Stem 7, 11, 13, 15, 16, 18, 22, 23, 26–29, 34–36, 38, 40, 41, 43, 46, 48–54, 58, 59, 67
Stigma 39, 50
Stipule 51
St John's Wort 52
  'Rowallane' 21, 52
Strawberry, Garden 62, 71
Style 16, 50
Sumach, Stag's Horn 60
Sunflower, 28, 66
  Common 84
  Stiff 28

T

Tanacetum parthenium 87
Taraxacum 26
Tare, Hairy 31, 37
Teasel, Wild 67
Tendril 31, 51
Tepal 40, 43
Trefoil, Bird's-foot 31, 83
  Lesser 37
Trifolium dubium 37
  repens 32
    'Purpurascens Quadrifolium' 32
Tube 16, 40, 42
Tulip 16, 18, 43, 71, 72, 85
Tulipa 43, 85
Twig 18, 64, 72, 78

U

Umbel 34, 42
Urtica dioica 30, 69

V

Verbena 16, 17, 84
  × *hybrida* 17, 84
Vetch* 31
Vetchling, Meadow* 31
Vicia 31
  hirsuta 31, 37
Vinca major 17
Vine, Wild 65
Viola × wittrockiana 49, 86
Virginia Creeper 65, 69

W

Wallflower, 44
  'Bowles's Mauve' 44
Wall Lettuce 15
Walnut* 15
Whorl 8, 45, 50, 52, 54, 56
*Wisteria** 62
Woodrush, Field 37

Y

Yorkshire Fog 37